Demystifying NFV in Carrier Networks: A Definitive Guide to Successful Migrations

Copyright © 2014 Ixia. All rights reserved.

This publication may not be copied, in whole or in part, without Ixia's consent. Ixia, the Ixia logo, and all Ixia brand names and product names in this document are either trademarks or registered trademarks of Ixia in the United States and/or other countries. All other trademarks belong to their respective owners. The information herein is furnished for informational use only, is subject to change by Ixia without notice, and should not be construed as a commitment by Ixia. Ixia assumes no responsibility or liability for any errors or inaccuracies contained in this publication.

Table of Contents

Preface ... 1
    Why Read This Book? ... 3
    About Ixia ... 3

Introduction: Everything Known is Unknown Again ... 6
    Validating Real Performance in a Virtual World ... 8

Chapter 1: What is NFV? ... 10
    1.1 A Move Toward Software, Simplicity, and Standardization ... 10
    1.2 The Market for NFV ... 17
    1.3 Vendor Milestones ... 18
    1.4 Carrier Milestones ... 20
    1.5 NFV in the Mobile Core ... 21
    1.6 Why Brave Another Paradigm Shift? ... 22

Chapter 2: Drivers for NFV ... 24
    2.1 Increased Service Agility and Flexibility ... 25
    2.2 Economic Advantages ... 27
    2.3 Customers Benefit as Well ... 28

Chapter 3: NFV Risks and Challenges ... 30
    3.1 Strategic Challenges ... 31
    3.2 Architectural/Implementation Challenges ... 35

Chapter 4: Validation Strategies ... 44
    4.1 Virtual Testing vs. Testing Virtualization ... 45
    4.2 Critical Components of Validation ... 48
    4.3 Key Performance Indicators (KPIs) ... 49
    4.4 Scope of Validation ... 50

Chapter 5: Evaluating the NFV Infrastructure — 54

- 5.1 Hardware — 54
- 5.2 The vSwitch — 55
- 5.3 The Hypervisor — 56
- 5.4 VM Manager — 57
- 5.5 Management and Orchestration — **58**
- 5.6 Virtual Machines (VMs) — 59
- 5.7 Real-world Scenarios — 59

Chapter 6: NFV Test Cases — 62

- 6.1 Basic Test Setup — 62
- 6.2 Validating the Virtual EPC — 66
- 6.3 Testing a Network with a Combined Legacy and Virtualized EPC — 68
- 6.4 Testing Virtualized Elements Within the EPC — 70
- 6.5 Testing IMS Virtualization — 71
- 6.6 Testing vBRAS Functionality — 73
- 6.7 Beyond Migration — 73

Chapter 7: Maintaining Visibility— Monitoring, Access, and Control in a Virtualized Environment — 76

- 7.1 Virtual Visibility: Evolving Goals and Best Practices — 79
- 7.2 Phantom Taps and the New Virtual Visibility Framework — 80

Chapter 8: Ixia NFV Solutions — 84

- 8.1 Comprehensive "Real" and Virtual Test Capabilities — 84
- 8.2 The Most Trusted Names in Networking Trust Ixia — 87

Acronyms and Terms — 88

Preface

Network Functions Virtualization (NFV) is one of the most profound paradigm shifts the networking industry has faced to date. Proven functions such as routing, policy, firewall, DPI, and many others will move from running on dedicated hardware appliances to running on unproven virtualized server platforms in the hope of achieving massive efficiencies.

While carriers agree on the need and vision for NFV, many are struggling to quantify the benefits, understand the practical migration steps, and measure success:

- How should they go about replacing their current network architecture, moving from purpose-built physical devices to a virtualized, software-controlled scenario?
- How will they know with certainty that the migration was successful and justifies their investment?
- Does the virtualized solution deliver the performance, resiliency, security, elasticity, and quality of experience (QoE) needed to satisfy customers?

With network usage, user expectations, and competitive threats all on the rise, the risks to operators' brands and bottom line are too great to leave NFV deployments to chance. Everything known becomes unknown again, and there are risks associated both with moving too fast, and not fast enough.

From beginning to end, targeted test strategies and methodologies are needed to accelerate and ensure the delivery of NFV solutions with guaranteed quality. Ixia solutions validate that the motivations for NFV are achievable prior to migration, and that the expected benefits are realized in the virtualized environment. We help operators understand when it makes sense to virtualize an aspect of the network and when it doesn't, delivering clear insight into the migration process and its ultimate success.

The promises of NFV are vast and enduring, but operators face many diverse challenges in ensuring the same — or better — performance, reliability, and security achieved by traditional infrastructures.

Why Read This Book?

Demystifying NFV helps fast-track planning and deployment by exploring:

▶ What NFV is and what changes

▶ Virtualization market dynamics and migration forecasts

▶ Benefits of adopting NFV

▶ Deployment challenges

▶ Evolving best practices for evaluating and validating NFV strategies and initiatives

▶ High-level test cases for NFV validation

We'll begin with a bit more background on what network functions virtualization is and the potential benefits it stands to deliver, then explore the obstacles to deployment and strategies for overcoming them.

About Ixia

Ixia is an active participant in the European Telecommunications Standards Institute (ETSI) Industry Specification Groups (ISG) for NFV, a chair on the Open Networking Foundation, and remains involved in other leading industry organizations focused on virtualization. We continue to work with equipment manufacturers and service providers at the forefront of virtualization to define best practices for a smooth and profitable migration.

As "paradigm shifts" become the norm for mobile operators, our unrivaled experience working with leading providers worldwide helps both in transforming networks and fast-tracking the delivery of compelling new services using virtualized infrastructures. Ixia's unique, comprehensive portfolio of hardware and virtualized test and monitoring solutions deliver the "lab-to-live" insights needed to accelerate and maximize the benefits of NFV throughout the deployment life-cycle.

Our leadership in virtualization performance testing and optimization equips service providers to approach and implement Software Defined Networking (SDN) and NFV with the greatest possible confidence, efficiency, and support.

Related eBooks From Ixia

For additional insight into major transformations occurring in service provider networks worldwide, explore other timely Ixia resources:

- *Validating VoLTE: A Definitive Guide to Successful Deployments*
- *Small Cells, Big Challenge: The Definitive Guide to Designing and Deploying HetNets*
- *Authoritative Guide to Advanced LTE Testing*

Introduction: Everything Known is Unknown Again

Driven by providers' ceaseless quest to deliver and monetize new services, mobile network infrastructures are evolving faster than at any time in history:

- ▶ Demand continues to grow exponentially
- ▶ All-IP infrastructures have become a reality
- ▶ Radio access networks are diversifying as HetNets emerge introducing small cells and carrier Wi-Fi
- ▶ 5G is on the radar screen

Amidst this rampant change, the most transformative shift underway is one designed to speed and facilitate change itself. Network Functions Virtualization (NFV),

> *"NFV, done correctly, will be transformational."*
> — **Heavy Reading**

basically the shift of network control to software to gain agility and a host of operational and economic advantages, is quickly moving forward.

In tandem with SDN, the migration of networking functions to a virtual or cloud-based architecture promises to play a pivotal role in operators' ability to innovate new services, accommodate growth, and compete profitably into the future.

Wide-ranging benefits include both technological and economic advantages:

- Accelerating the introduction of compelling new services
- Increasing network and service agility
- Simplifying network management and optimization
- Improving the economics of service delivery with extensive capex/opex savings

Proof-of-concepts (PoCs) are underway worldwide and analysts foresee NFV spending growing at a CAGR of 46% between 2014 and 2019.[1] But the risks are great, and the challenges formidable, and still somewhat unknown.

NFV effectively means having to create and manage a "carrier grade" cloud. And where the capabilities and performance of physical devices are well known and understood, virtualizing various network functions renders these aspects unknown once again.

Throughout the process, operators must:

- **Transform the unknown into the known** by validating new architectural components, assessing the performance of traditional functions getting moved around, and cultivating new teams, vendor relationships, and skill-sets
- **Adopt an infrastructure model developed for data centers** versus carrier infrastructures and requirements
- **Invest in and fast-track PoCs** while standards are still evolving
- **Leverage evolving best practices** to validate decisions, weigh potential benefits against tradeoffs, and measure success

---

1 http://www.mindcommerce.com/Publications/NFV_BusCaseMarketAnalysisForecasts_2014-2019.php

Validating Real Performance in a Virtual World

With NFV, the old approach to making decisions—relying heavily on performance data provided by manufacturers—is not only impractical but impossible because real-world data does not exist yet. In Chapters 4-7, we'll offer some strategies and methodologies for approaching and validating vital decisions including:

- ▶ Evaluating new devices and architectural models
- ▶ Replicating the complexities of virtualized wireless networks
- ▶ Measuring and guaranteeing subscriber quality of experience (QoE)
- ▶ Simulating security attacks
- ▶ Monitoring live virtual networks

"Lab-to-live" strategies are needed to quantify and evaluate performance, move forward with confidence, and maintain quality as traditional and virtualized networks coexist for the foreseeable future. But let's take a step back and briefly summarize what NFV is, why operators want it, and the obstacles they face in making it happen.

What is NFV?

Chapter 1: What is NFV?

In this chapter we'll take a brief look at what Network Functions Virtualization (NFV) is, its relationship to Software Defined Networking (SDN), projected growth, and early efforts by network equipment manufacturers (NEMs) and operators to date.

1.1 A Move Toward Software, Simplicity, and Standardization

As is commonly known, the NFV concept was first formally outlined in October 2012 in a white paper published by an ETSI industry specifications group. The Network Functions Virtualization ISG that initially consisted of representatives from thirteen service providers worldwide has grown steadily in size and scope. More than 150 organizations now participate, and several subsequent papers have addressed NFV use cases and implementation.

Conceptually, NFV marks a fundamental shift in the service provider network model in which diverse networking functions move to a virtual or cloud-based architecture. In practice, NFV may entail:

- ▶ Migrating networking functions from proprietary, specialized hardware appliances to commercial off-the-shelf (COTS) x86-based servers

- ▶ Abstracting and shifting control of networking functions from hardware to software by introducing a hypervisor layer

- ▶ Flexibly distributing functionality across desired locations — data centers, network nodes, customer premises—to maximize operational efficiency and performance

- ▶ Creating a more "application-aware" network

Targets for NFV include everything from switching, routing, and broadband remote access servers (BRAS) to access devices to load balancing and critical security elements such as firewalls and deep packet inspection (DPI). In mobile networks, NFV represents an end-to-end proposition reaching from the IP multimedia subsystem (IMS) and evolved packet core (EPC) to radio access networks (RANs) to the customer premise.

While virtualization will likely occur on a data center by data center basis initially, the ultimate vision allows for virtualized network functions (VNFs) to run in the cloud over logical and geographically dispersed topologies to achieve maximum efficiency and the lowest possible total cost of ownership (TCO). The figure below depicts the evolving NFV architecture.

High-Level NFV Framework

The proposal is for VNFs to run on COTS hardware such as Dell/HP servers with standard x86-based computing architectures. The new **virtualization layer** shown in the figure above is a hypervisor that provides virtual access to underlying compute resources, enabling features like fast start/stop of virtual machines (VMs), snapshot, and VM migration.

Hypervisor software is able to manage several guest operating systems and enable consolidation of physical servers onto a virtual stack on a single server. CPU, RAM, and storage are flexibly allocated to each VM via software.

Management and orchestration functions will undergo a profound transformation as this layer must interact with both virtualized server and network infrastructures, often using OpenStack protocols, and in many cases SDN. This highest layer is referred to as the Operations Support System/Business Support System (OSS/BSS), a critical component of, and perceived obstacle to the widespread adoption of NFV.

Ultimately, cloud-based management and orchestration stands to facilitate management of large, highly-distributed network infrastructures and innovative service offerings. The ETSI ISG is also proposing an architecture for **Service Chaining** whereby multiple network functions are strung together, typically interconnected by a vSwitch.

A virtual network function forwarding graph, or VNF FG can be created, scaled, and updated very quickly and efficiently. Virtualized functionality can be added to service chains by instantiating a VM and simply updating the forwarding graph.

As shown in the diagram below, functions can be nested instead of linear, and stitched together from different physical locations using various virtualization solutions such as overlay (VXLAN, NVGRE, STT) and SDN/OpenFlow.

End-to-End Network Service

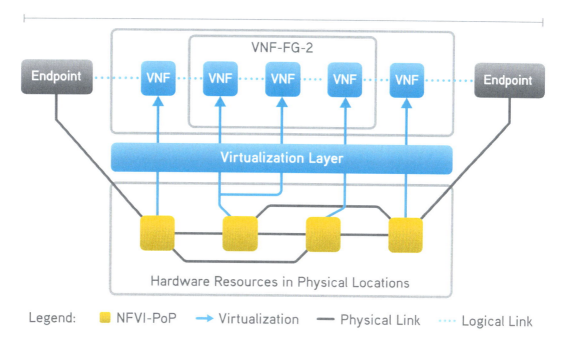

Example of an end-to-end network service with VNFs and nested forwarding graphs

With NFV, operators gain greater flexibility in defining service chains such that a particular service follows the path most applicable to it.

1.1.1 NFV and SDN

In honing their virtualization strategies, operators will likely evaluate and leverage NFV and SDN in tandem. SDN enables the decoupling of the control plane from the data plane, offering increased programmability. The ability to program the network through software in turn promises simplified traffic management and greater ability to define the way in which packets are forwarded by networking elements.

The shift away from distributed protocols such as Border Gateway Protocol (BGP) and Open Shortest Path First (OSPF) toward more centralized control (OpenFlow is the leading SDN protocol thus far with support by more than twenty leading vendors) allows various types of equipment from multiple vendors to be more easily monitored and managed. As is a goal with NFV, the more granular control enabled by SDN allows services to be created and deployed across networks with reduced equipment requirements. Increased visibility further equips operators to deliver a higher quality of experience.

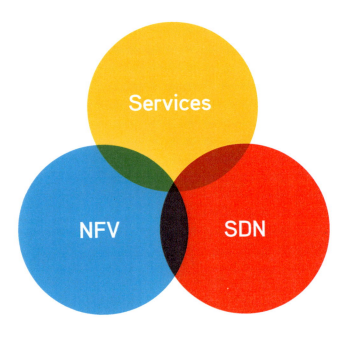

Industry experts resoundingly agree that NFV and SDN are highly complementary strategies, though services can be built directly using NFV without SDN, or be built using SDN without migrating functions to the cloud. Combining the two stands to deliver compounded benefits, however, as SDN can be used to provision network connectivity to VNFs, allowing end-to-end services to be built with enhanced virtualized functions.

Several top use cases for OpenFlow-enabled SDN and NFV can be seen in the Open Networking Foundation document "OpenFlow-enabled SDN and Network Functions Virtualization" (February 17.2014).[1]

Big Picture View of NFV with SDN and Management

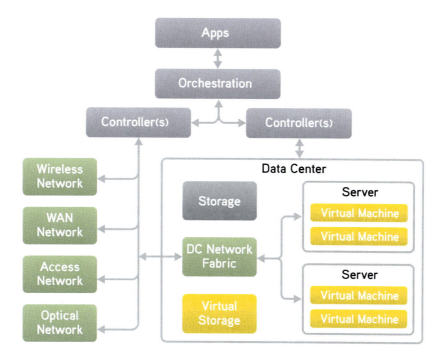

Source: Alcatel-Lucent[2]

1 https://www.opennetworking.org/images/stories/downloads/sdn-resources/solution-briefs/sb-sdn-nvf-solution.pdf.
2 "Network Functions Virtualizations: Challenges and Solutions," Alcatel-Lucent, 2013.

1.1.2 The Evolution of NFV

Late last year, an industry-wide call was made for public PoC demonstrations with some twenty-five providers banding together to define progress and processes. Since then, several have embarked on PoCs.

PoCs are organized by the ETSI ISG for NFV and must include a service provider and two or more NEMs. Eighteen have been initiated to date.

Visions for the adoption of NFV vary, as individual operators' plans undoubtedly will. In the diagram below, "Virtualized" network functions are those running on hypervisors and general-purpose hardware but potentially requiring dedicated physical resources and customized configurations. The "Cloud" refers to virtualization-based functions deployed on standard interfaces, while "Automated Life-cycle Management" refers to the eventual scenario in which service providers use tools similar to those found in the IT world to manage the VNF life-cycle.

NFV Deployment Evolution

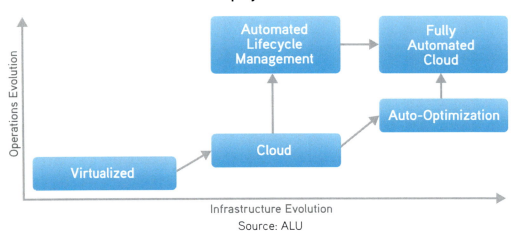

Source: ALU

As described by ALU, "Auto-optimization" describes the stage in which VNFs are able to dynamically and automatically scale to match available resources with changing demand.

1.2 The Market for NFV

As typically occurs in nascent markets, growth forecasts are all over the map:

- Mind Commerce estimates global spending on NFV solutions will grow at a CAGR of 46% between 2014 and 2019. NFV revenues will reach $1.3 billion by the end of 2019[3]
- The Dell'Oro Group thinks the market could represent $2 billion in equipment sales by 2018[4]
- Doyle Research envisions a "best-case" scenario in which NFV gains rapid momentum reaching a $5 billion market by 2018 (including software, servers, and storage)[5]

As always, it's a matter of what aspects of the process each unique projection encompasses, and how quickly the industry can overcome the challenges described below in Chapter 3. Of greater and more immediate importance to operators is how NFV itself will evolve.

This year, Infonetics believes many operators will move from PoCs to collaborating with vendors to develop and produce software solutions that will furnish a foundation for commercial deployments.[6]

---

[3] www.mindcommerce.com/Publications/NFV_BusCaseMarketAnalysisForecasts_2014-2019.php
[4] www.sdncentral.com/news/nfv-market-size-2b-first-guess/2014/04/
[5] www.lightreading.com/forecasting-the-nfv-opportunity/a/d-id/705403
[6] "2014 SDN and NFV Strategies: Global Service Provider Survey." Survey of worldwide service providers controlling more than 50% of global telecom capex and 47% of revenue

1.3 Vendor Milestones

New and traditional vendors are bringing diverse architectures and products to market:

- ***Alcatel-Lucent*** has stated that it's working with some 20 service providers on NFV-related initiatives. ALU launched a portfolio of virtualized mobile network function applications spanning the EPC, IMS, and radio access network (RAN).

  The provider says several network functions are already available in a virtualized form, with others targeted near-term. ALU has also stated plans to launch a consultancy practice aimed at helping operators navigate the transition.

- ***Broadcom*** has also been aggressive, working with ARM to develop a server-class, 64-bit system on a chip (SoC) optimized for NFV that may ship in volume in 2015. Its Open NFV platform leverages open-source components such as Linux, KVM virtualization, and Open Virtual Switch to create a platform that is not reliant on a single-chip architecture in order to increase portability and enable the creation of NFV applications that are highly portable from one platform to another.

- ***Cisco*** has been showcasing its Evolved Services Platform, a unified virtualization and orchestration software platform designed to equip operators to automate and provision services across compute, storage, and network functions in real time. According to Cisco, the platform facilitates virtualization across a carrier's enterprise architecture, including cloud, video, mobile, and fixed networks.

- ▶ ***Hewlett-Packard*** recently launched an open, standards-based reference architecture. OpenNFV aims to provide a complete architectural approach across servers, storage, and networking to reduce costs. The company has launched a business unit dedicated to NFV, and introduced OpenNFV Labs and an OpenNFV Partner Program to speed development of NFV-based applications.

- ▶ ***Huawei*** has proclaimed NFV a main focus for the coming years with offerings including the CloudEdge solution designed for mobile broadband networks (MBB). CloudEdge purports to help operators automate network management functions and speed time-to-market for new machine-to-machine, mobile video, and mobile enterprise applications.

- ▶ ***Intel*** has teamed up with Red Hat on an ETSI-approved PoC for a virtualized EPC.

- ▶ ***Juniper*** unveiled a suite of SDN and NFV technologies designed to complement its Contrail SDN controller at Mobile World Congress 2014. Its NorthStar SDN controller also targets new network management software for automating the control of routers and other networking devices, as well as optical and mobile devices, from a single management plane.

- ▶ ***NEC's*** vEPC solution targets two core network functions: vMMEs (virtual Mobility Management Entities) and vS/P GWs (serving and PDN gateways). Virtual network functions are supposedly decomposed into elementary virtual machines.

- ▶ ***Wind River*** launched Carrier Grade Communications Server, an NFV platform intended to help carriers migrate their existing networks to NFV architectures with minimal disruption and leave them better able to deploy applications "out of the box."

1.4 Carrier Milestones

Infonetics has called 2013 the year of the PoC with operators, manufacturers, and software providers all beginning to wrestle with deployment issues such as complexity, cost, benefits, and tradeoffs in their labs. Early efforts have demonstrated some benefits with regard to scaling traffic and the allocation of resources in accordance with demand.

This year, the firm expects serious lab trials to progress to aggressive field trials with commercial deployments debuting in earnest in 2015 and becoming widespread by 2016. Among the early leaders are:

- ***AT&T's*** launch of the next generation of its Supplier Domain Program – Domain 2.0 – in 2013, targeting migration to modern, cloud-based architectures. Calling the approach a "transformative initiative," AT&T expects the project to leverage both NFV and SDN to accelerate time-to-market for advanced products and services. Initiatives target virtualizing the EPC to create a multi-service "user-defined network cloud" supporting a wide range of network functions and services.

- ***British Telecom (BT)*** co-founded and remains active in the ETSI NFV ISG and claims to be the first network operator to publish PoC results. BT has engaged in several NFV PoCs with various partners. Projects to date include:
  - Testing and successfully deploying a virtualized BRAS (vBRAS) solution
  - Testing and deploying virtualization of IPSec tunnel termination for Wi-Fi and LTE services
  - Testing virtualizing content distribution networks (CDNs)

- **Colt,** another original member of the NFV ISG, has several projects underway to leverage NFV and SDN including L3 CPE virtualization and WAN SDN in the data center. External projects include "Colt Live" targeting the media sector, "elastic bandwidth," and a "Service Console and Dashboard" for enhanced self-service and reporting.

- **Deutsche Telecom (DT)** has embarked on testing and deploying its Terastream Architecture in Croatia (based on NFV and SDN), which includes virtualizing DHCP, IPv4, and VPNs. The company has also tested distributing BRAS/MBG functionality performing authentication with SDN, and is discussing migrating the set-top-box to the cloud.

- **NTT DOCOMO** has announced plans to commercially deploy services using a virtualized mobile network during 2016. The carrier began collaborating with ALU, Cisco, and NEC on NFV in 2013 and recently announced the completion of PoC trials verifying feasibility of a virtualized EPC.

1.5 NFV in the Mobile Core

With regard to the evolution of NFV in general, a recent white paper published by Heavy Reading leveraging data from Radisys predicts that "Phase I" of the NFV/SDN transformation will be completed within the next five years.

In terms of the mobile core specifically, Infonetics asked operators about initial target applications. Fifty-nine (59%) of those responding for NFV reported plans to deploy mobile core/vEPC by 2016 or later, and many reportedly plan to leverage a vIMS core for VoLTE deployment as well.

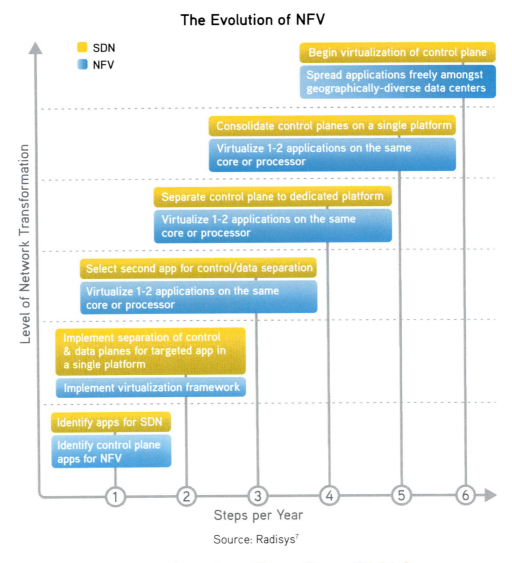

Source: Radisys[7]

1.6 Why Brave Another Paradigm Shift?

However fluid and subjective growth forecasts might be, NFV appears to be real and imminent, though it also looks to be a whole of work. For all three to be true at once, the prospective benefits need to be near-term, and far-reaching.

We'll take a look at the drivers for virtualization first, then at the challenges, and evolving best practices for making sure it works.

---

[7] "Software-centric Networks: A Migration Path to NFV," Heavy Reading, October 2013

Drivers for NFV

Chapter 2: Drivers for NFV

NFV and SDN figure to preserve and grow the profitability of services into the future. The industry is essentially taking a page from the IT world, where the virtualization of servers has enabled:

- ▶ **Greater service agility and programmability** including the ability to program the delivery of services through open APIs
- ▶ **Increased scalability and elasticity** with the cloud-based model enabling infrastructure as a service (IaaS), platform as a service (PaaS), and software as a service (SaaS) offerings
- ▶ **Cost-savings** through the consolidation and improved management of servers while enabling a "pay-as-you-go" model

In virtualizing networking functions, operators are seeking essentially these same benefits. The presumption is that, de-coupled from the hardware layer, telecommunications network applications can run on lower-cost, standards-based hardware with far greater flexibility and efficiency.

Executed correctly, NFV represents greater ability to roll out new, more compelling and profitable services faster, and with guaranteed quality. Increased efficiency and savings figure to prove even more valuable in mobile networks where margins have historically been tightest and competitive threats from over-the-top (OTT) providers continue to mount.

The virtualization of network functionality stands to deliver two basic benefits: Increased service agility and substantial cost-savings. Each involves many components, representing many long-term advantages.

2.1   Increased Service Agility and Flexibility

Ultimately, service agility translates into the ability to innovate and launch services faster and more cost-effectively. The evolution from proprietary equipment to a software-based environment:

- ▶ *Simplifies provisioning,* allowing operators to more quickly instantiate, move, and evolve services.

- ▶ *Adds network elasticity,* increasing their ability to scale up and down to meet demand.

- ▶ *Increases automation*, which in turn simplifies operations. Provisioning for virtual appliances needs to be automated in order to address the dynamic NFV environment. Doing so reduces provisioning and configuration times along with manually induced configuration errors.

- ▶ *Enables more fluid and efficient resource allocation*. Working in software facilitates specification of CPU, memory, and other resources, allowing providers to vary allocation and use according to specific use cases and changing needs. This in turn enables more fluid capacity management to accommodate flux in demand, failover, and the like. Higher resource utilization also results in less equipment being required.

- ▶ *Requires less specialization* as operators can increasingly leverage tools and skill-sets widely found in IT.

▶ *Allows greater proximity to popular IP services.*
Virtualizing networking functions allows services to be provisioned closer to cloud-based services like Facebook or Pandora, potentially enabling better performance and economic benefits.

NFV Deployment Drivers

Drivers	%
Scale services up or down quickly	86%
Use software for quick revenue	69%
Use commercial servers, not network equipment	62%
Operational efficiencies	59%
Multi-tenancy	45%
Real-time network optimization	34%
Save energy consolidating workloads	28%
VNFs from small players	14%

Percent of NFV Respondents Rating 6 or 7
Source: Infonetics[1]

Flexibility extends to certain aspects of marketing as well. As the Infonetics study discusses, providers can use both SDN and NFV to cost-effectively test new services before committing to full-blown rollouts. The ability to quickly tweak offerings based on early user feedback helps in bringing more compelling and profitable services to market faster, with a lower upfront investment and greater confidence that they'll succeed.

---

1 "2014 SDN and NFV Strategies: Global Service Provider Survey." Survey of worldwide service providers controlling more than 50% of global telecom capex and 47% of revenue

2.2 Economic Advantages

Predictably, another pervasive driver of virtualization is its potential to deliver higher return on investment (ROI) near-term, and lower TCO long-term. NFV's economic advantages include significant reductions in both capital and operating expenditures.

2.2.1 Capex Reduction

The most obvious capital savings stems from trading costly proprietary hardware for lower-priced commercial platforms. A less obvious but far greater benefit is not having to invest as heavily in redundant backup devices—MMEs, SGWs, PGWs, and the like—that largely go unused because of forced redundant configurations. Migration to a distributed cloud architecture allows operators to deploy backup in software using an N+1 configuration, versus having to invest in and warehouse lots of extra equipment.

2.2.2 Opex Reduction

While the capex savings may feel more immediate, a far greater benefit of NFV is the potential for pervasive and compelling savings on operating expenses achieved through increased efficiencies:

- ▶ The ability to share computing resources between functions using hypervisor technology
- ▶ Reduced power, space, cooling requirements
- ▶ Use of widely available tools and skill-sets
- ▶ Reduced management costs through increased automation and more efficient use of resources
- ▶ Reduced field upgrades (truck rolls) required than with proprietary hardware

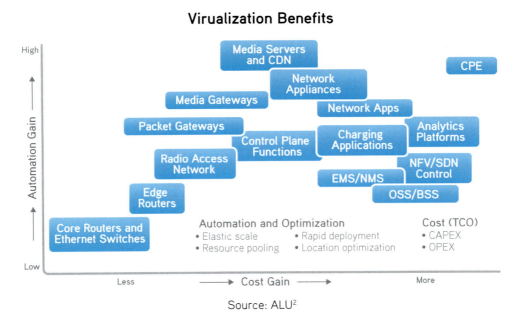

Source: ALU[2]

Virtualization stands to play a key role in helping operators predict and manage costs as networks scale to accommodate exploding demand for new services. According to the Infonetics survey, operators see virtual enterprise CPE, or vE-CPE, as the top use case for NFV opex benefits by virtue of increased service agility to businesses. Mobile core/vEPC comes second, followed by service chaining and a virtualized IMS core (vIMS core), all of which are of particular interest to mobile operators.

2.3 Customers Benefit as Well

As operators simplify operations and potentially reduce cost, benefits may extend to the end-user as well. NFV will help to enable features such as self-service portals that allow customers to change services in real-time. An attractive "pay-as-you-go" model may also arise versus requiring users to purchase capacity for future needs.

---

2 Network Functions Virtualizations: Challenges and Solutions," Alcatel-Lucent, 2013.

NFV Risks and Challenges

Chapter 3: NFV Risks and Challenges

Amidst exponential traffic growth and rising user expectations, NFV essentially calls for the creation of a "carrier grade cloud," an unknown quantity to say the least. The biggest risks and challenges inherent in virtualization center on guaranteeing quality and performance.

Deploying latency-sensitive networking applications in the cloud may impact performance in unforeseen or unacceptable ways, or limit the ability to react to changes in usage profiles without incurring degradation. Service level agreements (SLAs), QoS requirements, and expectations for quality are already in place, and operators can't afford to let them slip.

The same is true of reliability, availability, and security. Overall performance and satisfaction must remain as good as they are today, or get even better as virtualized environments scale in size, scope, and complexity.

NFV standardization efforts continue, but more work, and proof, is needed to fully inspire confidence. Operators are moving forward anyway as we've seen, tackling strategic and technological challenges in tandem.

The bottom line is: jobs are on the line with NFV, with huge potential downsides both for moving too fast and not getting it right, or moving too slow, and not getting it done.

3.1 Strategic Challenges

In embarking on virtualization, operators need to establish a business case, and a plan for determining what to virtualize and when. They then must determine how best to approach migration, and establish a means for demonstrating it worked.

3.1.1   Challenge 1: What to Virtualize and When

Determining the value of NFV is essentially a game of evaluating the tradeoffs between openness and performance, flexibility and control, and quality and cost. At the highest level, operators may also need to decide between an application-driven strategy where they essentially test NFV out on a particular function, and a more aggressive, platform-driven strategy designed to achieve substantial virtualization much more quickly and cohesively.

Functions that are mostly control-plane-centric are great first candidates for virtualization. In mobile networks, certain elements in the EPC and IMS networks make ideal candidates to virtualize. These include the MME, all Diameter servers (HSS, PCRF, OFCS and OCS), and x-CSCFs (call session control functions). While all of these have real-time performance requirements, they do not perform any user-plane processing, and therefore don't have the same criticality as devices such as SGWs and PGWs on the user plane. CPE and broadband access devices are other strong candidates.

In some cases, the benefits of virtualization are pretty clear. Other elements are obviously less ideal candidates, such as those with challenging real-time requirements. For example, the potential gains achieved by virtualizing data-plane routers, Ethernet switches, layer 1 processing functions, and encryption may not stack up against the cost, deployment effort, and potential performance tradeoffs. That said, some packet processing applications are now running on more standard compute platforms, and others moving to COTS if not yet to fully virtualized platforms.

Between the two extremes—obviously good candidates for NFV and clearly bad candidates—many networking functions and devices fall somewhere in the middle, with a host of variables coming into play. By way of example, elements with strict real-time requirements, or that use custom ASICs, may be more challenging and require more server blades, but still merit consideration due to long-term efficiency gains.

For some functions, the benefits will be greater in certain deployment scenarios than in others. In evaluating candidates for NFV, a wide range of qualitative and quantitative criteria and dependencies may come into play:

- Can services run together effectively in the same server?
    - If no substantial savings result from virtualizing a function, are gains achieved through automation enough to make it a good candidate for NFV anyway?
- How will it scale?
    - Scalability can be measured and achieved in many ways. Will a single VNF be used to replace an appliance, or will many VNFs be used or distributed? This can also impact service design.

3.1.2 Challenge 2: How to Go About It

Virtualization will typically occur in stages. This may delay the ultimate benefits, but will afford operators the chance to evaluate the risks and benefits during each phase.

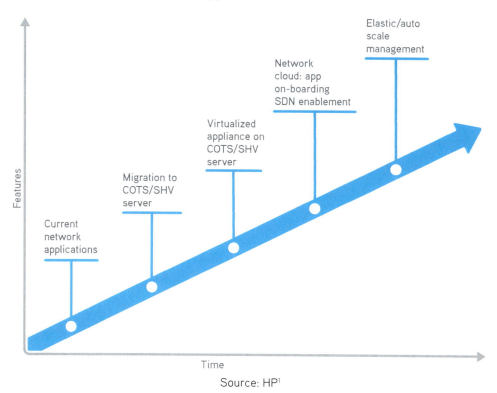

When moving functions from physical appliances to virtual instances, there can be many trade-offs and many design decisions that need to be made:

▶ How should server resources like memory and CPU be allocated to each function?

▶ Should resources be optimized for specific applications?

---

1 "Technical White Paper: Network Functions Virtualization," HP, 2014

- How should NFV be integrated with what exists today?

- Should a contained or hybrid approach be used?

- Where will the VNF be deployed? Is there a regional data center? Will the location of the VNF impact the service delivery due to the network delay?

One challenge may arise from recent centralization of data centers. For example, if a customer is located in California and the nearest data center is in Colorado, operators must evaluate whether it makes sense to provide a network function in the CO data center and risk inducing latency.

Service providers must validate decisions and "what if" scenarios each step of the way, asking other key questions such as:

- *"How much, how soon?"* The economic benefits of NFV will likely increase as initiatives scale, but aggressive virtualization means a profound transformation of existing paradigms. Most operators will likely tackle the shift in phases, which means having to define the various stages, and determine which elements will be addressed within each, and in what order and combination.

- *"Who should do what?"* NFV requires a combination of extensive network and data center expertise. Operators must decide:
    - Which members of which teams should be involved
    - Whether new or additional expertise is required
    - If and when it makes sense to outsource design, testing, or integration

3.1.3 Challenge 3: Measuring Success

"How do we know it worked?"

In theory, the ability to meet SLAs and deliver reliable QoE to subscribers shouldn't change when SDN and NFV are applied. And once again, the flexibility achieved through virtualization should ultimately result in quality changing for the better.

Measuring the success of virtualization is really the same as measuring the performance of a dedicated hardware-based system. The same KPIs apply with virtualization as before, and the two will be compared against one another.

For NFV to be considered a success, the performance of the virtualized system must be at least equal to that of the legacy system. And as might be expected, problems may arise more frequently in the early stages of adoption requiring different aspects of performance to be evaluated as the various kinks get worked out.

Goals must be clearly defined and reliable strategies in place for validating and improving performance from start to finish.

3.2 Architectural/Implementation Challenges

The primary architectural benefits of virtualization—elasticity, nimbleness, and openness—represent challenges in and of themselves. In achieving them, operators will need to dramatically increase automation and real-time control of both physical and virtual resources.

Management, security, and visibility strategies must also become flexible and adaptable enough to address hybrid environments that encompass both legacy and newly virtualized functions. High-level architectural and performance challenges include:

3.2.1 New Devices, Techniques, and Dependencies

While the ultimate goal of NFV is simplicity, the process of virtualizing new and traditional functionalities adds new degrees of complexity:

- Will the IT model work? Can proven, reliable elements move from custom hardware to unproven, and even untested software-based solutions without performance suffering?

- Will combining particular functions on a single blade or processor impact performance?

- Will bringing multiple elements like MMEs, SGWs, and IMS core elements together degrade the performance of individual components?

The adoption of NFV necessitates measuring the performance, scale, and interdependencies of newly virtualized elements, as well as the performance of the overall architecture end-to-end.

3.2.2 Speed

Will software be fast enough even with continuing advances in CPU technology? The new Intel Xeon E5 v2 reportedly achieves speeds up to 250Mpps, theoretically sufficient for most networking applications, but providers need to be sure it all works in practice. Having the right Network Interface Cards (NICs) within the server also plays a role, especially while moving from 1G to 10G to 40G and maintaining multiple NICs.

In addition to leveraging generic hardware as a platform, other considerations will impact the performance of VNFs, mainly the fact that there will likely be other VNFs running on the same hardware. Multi-tenancy can introduce a degree of unpredictability into the expected performance of each newly virtualized function.

3.2.3   Service Delivery

Virtualization should not impact existing services. The underlying delivery mechanism, whether it be virtualized or not, should be completely transparent to the services being delivered by the network.

This is critical since some network functions will likely be virtualized in stages with other parts continuing to be implemented on a traditional hardware-specific infrastructure. Throughout the process, the availability of services should meet the same high standards that are in place today.

3.2.3.1 Service Chaining

As noted above, one advantage of NFV is allowing more flexible service chaining, or VNF forwarding graphs. In the traditional model, most services had to follow the same path, whether there was value in having them pass through specific functions (DPI, firewall, IPS) or not.

With NFV, the operator has greater flexibility to define specific service chains for specific services, such that a service only follows the path that is directly applicable to it. However, in implementing VNF FGs, operators must be able to guarantee sufficient capacity and resiliency while increasingly automating provisioning.

These changes in the service chain should be tested prior to deployment to ensure that changes do not impact service delivery in unexpected ways.

3.2.4 Increasingly Distributed Infrastructure

Virtualized networks will become more geographically distributed over time, making it harder to predict and control latency and other variables. One of the benefits of NFV is the ability to quickly instantiate systems, but sometimes the motivation behind this could be the localization of services. This will increase the distributed nature of the network itself, which can be taken as both an opportunity (to reduce response latency times to the user) and a challenge (latency times required to communicate with the home system may need to be increased).

3.2.5 Multi-tenancy

Multi-tenancy becomes a challenge as the cloud-based approach evolves. Operators must be able to manage policies for individual services and flows as functions are decoupled from physical devices.

3.2.6 Scalability

NFV needs to be massively scalable to support large numbers of data centers and millions of subscribers. In addition, the scalability of resources will be far more dynamic in a virtualized environment.

The major advantage here is elasticity as VNFs can be created, adjusted, and destroyed in real time, and on demand. When network triggers are reached, capacity can be dynamically added or removed from the overall network such that capacity and performance can constantly change to reflect the current demand.

Networks must be capable of being reconfigured rapidly to achieve the desired elasticity. Virtualized functions will need to be dynamically updated as a result of scaling resources. For example, the domain name system (DNS) service, which can be responsible for load balancing between VNFs providing the same service, must be instantly made aware of new elements being brought online as a result of scaling.

3.2.7 Management and Orchestration

Operators must devise and implement new approaches for managing virtual functions and networks. As part of this effort, 69% of respondents in the Infonetics survey cited OSS/BSS concerns as the biggest barrier to deploying NFV.

Operators need to figure out how to address virtualization across complex back office operations and business support systems. Ideally, the management of the virtualized elements will be transparent relative to the non-virtualized elements. Thus it should not matter to the management system if a function is virtualized or not.

3.2.8 Visibility

In a virtual environment, real-time visibility into the end-to-end architecture becomes even more critical to guaranteeing service availability and QoE. When things happen unexpectedly in either the test lab or the live network, operators can leverage virtualized taps (vTaps) and other monitoring tools to find bottlenecks, pinpoint performance issues, and test varying configurations.

NFV and SDN will give rise to new monitoring challenges and strategies requiring new visibility architectures and performance metrics for components such as hypervisors and VMs, as well as the performance of VNFs themselves.

We'll take a closer look at evolving strategies for visibility in Chapter 7.

3.2.9 Robustness

Robustness is a broad challenge that refers to the ability of the virtualized network to perform fault detection and invoke the associated diagnosis and recovery mechanisms if a fault is detected. This also means that the entire state of the faulty VNF must be maintained as it is transferred and recovered elsewhere.

For example, if a particular VNF is found to be having problems, a new instance of that VNF could be instantiated elsewhere in order to replace it, but all the sessions within the first VNF would have to be transferred to the new one. In some cases, it may be necessary to fall back to a non-virtualized function from a faulty virtualized function. This should be supported transparently.

3.2.10 Security

While virtualizing firewalls and load balancing functionality may enable more flexible, nimble defenses, NFV also introduces new elements, like hypervisors, that represent new targets for attackers. The more nebulous boundaries of the cloud and increased distribution of VMs across multiple geographic locations further complicate defense strategies.

New configurations, devices, and defense strategies must be validated prior to implementation and as networks evolve and scale up and down.

3.2.11 Measuring Performance

A new approach to performance validation is required to fully assess virtualization and the success of migrations end-to-end. The next two chapters take a detailed look at evolving test architectures, and real-world use cases.

Validation Strategies

Chapter 4: Validation Strategies

Perhaps the greatest challenge operators face in undertaking NFV and SDN is that getting started essentially means starting over. Everything old, known, and proven, becomes new, unknown, and unproven again.

In the process of becoming a VNF, tried-and-true networking elements must be re-evaluated from the ground up with an eye toward determining 1) Whether the business case supports migration, and 2) How performance might be impacted.

Operators need reliable tools and strategies for quantifying the benefits, costs, and risks associated with virtualizing individual elements and systems, along with a means of determining whether the process works and meets goals. New test capabilities and strategies must be introduced, and the same rigorous performance criteria applied.

Comprehensive NFV validation requires measuring the performance of newly virtualized networking functions, as well as the ability to assess the performance of the overall architecture and services end-to-end. For the greatest efficiencies, a combination of traditional and virtualized testing can be conducted, leveraging each approach where it makes the most sense.

How Do You Get From Here to There?

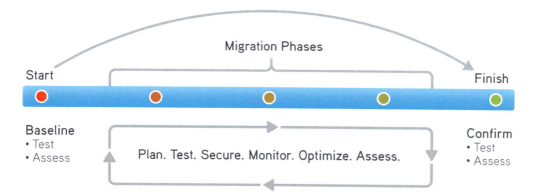

4.1 Virtual Testing vs. Testing Virtualization

In tandem with server virtualization efforts, leading test solution providers have introduced virtual versions or extensions of traditional hardware-based capabilities.

4.1.1 Benefits of Virtual Testing

Virtual test solutions introduce several critical advantages such as:

- The ability to quickly recreate environments
- Rapid reproduction of development, Quality Assurance (QA), and production environments through snapshots
- Automated creation and management of complete networks
- Validation of NFV configurations without "truck rolls"
- Remote debugging/reduced debug times

4.1.2 Which to Use When

The decision to use virtualized testing functionality versus traditional hardware-based test systems depends on goals and requirements, as well as the preference of the user. There are no clear-cut rules here, but several considerations factor into the decision, and each approach has its advantages for certain types and stages of testing:

- ***Performance testing.*** While virtualized testing appliances can easily be used for performance testing, traditional hardware-based systems have the advantage of known performance expectations. These expectations are validated and dimensioned under various conditions such that users know exactly what they're getting. In such cases, using traditional systems simply removes a variable from the testing process.

- ***Precise measurements.*** When measurements with a high degree of precision are required, typically with a resolution of less than 1ms, then using dedicated hardware for the test equipment is recommended. Specialized hardware will have an advantage over generic servers for precise time-stamping of the incoming and outgoing packets.

- ***Development testing.*** While bringing up a new service or network element for the first time, operators may wish to run single-shot functional tests. Here, the convenience of being able to simply start a VM with a test tool instance can be very powerful.

- ***Resource contention.*** When multi-person teams are performing testing, having the ability to simply "create" your own testing resource by instantiating it also proves powerful. In this case, virtual test appliances can be quite beneficial.

▶ **Test tool orchestration.** Where there is a requirement for the test tool to be managed and orchestrated by the same system as the VNFs under test, using virtualized testing appliances can provide a clear advantage.

▶ **Deployment environment testing.** This is a very unique capability provided by a virtualized test tool, where the tool itself can also be deployed in the cloud along with VNFs such that tests can be run before deploying the virtualized functions. Testing can be done post-deployment as well.

▶ **Visibility.** As networks will likely encompass both traditional and virtualized network functions for some time, virtual monitoring technology must also be added. Virtual taps help pinpoint performance issues both in the lab and in the field, demystifying the complexities and obscurity of newly virtualized environments.

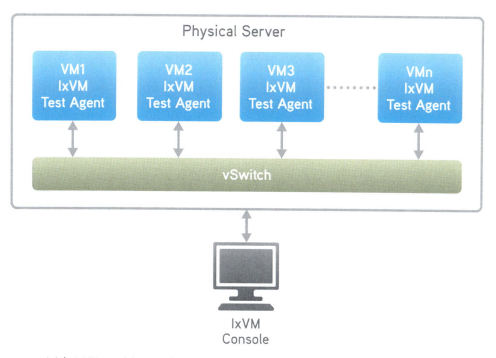

Ixia's IxVM provides a software-based version of traditional hardware test ports

Chapter 4: Validation Strategies

With so much occurring within a single server, it becomes increasingly difficult to isolate the potential sources of performance issues. This inability can result in finger-pointing and costly delays.

As we'll see in Chapter 7, a virtual visibility architecture is needed to isolate issues to new elements such as the vSwitch, hypervisor, or specific VNFs.

Standard Test Solutions	Virtualization Test Solution
Hardware-based traffic-generation / load testing	Software-based traffic-generation / load testing
Suitable for all performance, scale and system tests	Ideal for development and functional testing in virtualized environments
High-precision performance measurements	Low-latency precision and accuracy. Unlimited performance measurement platform

4.1.3 Requirements for Both Approaches

It will become increasingly critical to leverage both traditional and emerging virtual test capabilities. It will also be beneficial to have the two platforms interoperate and deliver the same sets of measurements.

4.2 Critical Components of Validation

Testing a virtualized implementation of the network requires the same important capabilities used in testing a dedicated implementation of the same network. Testing must be able to validate, for example, that the EPC and IMS are capable of delivering services with the same QoS, whether virtualized or not.

Assessing a virtual system requires measuring the various things that can have an impact on performance. This includes assessing the resources allocated to the VM (CPU, memory), ensuring against overprovisioning, and allocating the best resources for the expected performance. Critical capabilities should be validated as usual:

- User-plane performance (how much traffic can be carried through the network without QoS degradation)
- Capacity
- Control-plane performance
- End-to-end QoS
- Service validation
- Critical support features (charging and policy, DNS, DRA, security, etc.)

4.3 Key Performance Indicators (KPIs)

In evaluating the general performance of mobile core functions, typical KPIs include:

- Total user-plane throughput over default bearers (best-effort service)
- Total number of subscribers supported
- Total number of control-plane procedures per second supported (includes procedures such as attach, detach, service request, handover, IDLE mode to ACTIVE mode transitions, and the like)
- Session establishment availability and latency
- Dedicated bearer establishment availability and latency
- Handover latency

For QoE, related KPIs include:

- For best-effort, TCP-based services: time to first byte, time to last byte, TCP retransmissions, TCP resets

- For conversational traffic, typically over dedicated bearers: dropped packets, jitter, Mean Opinion Score (MOS)

- For VoLTE control plane: call establishment time, media cut-through after answer

When dedicated bearers are used, other typical quality of service (QoS) tests measure the performance of the guaranteed QoS (dedicated bearer) traffic when the amount of best-effort traffic is increased. In this case, the dedicated bearer traffic should have a relatively unaffected QoS, even while the best-effort traffic is increased.

Also see Ixia's *Authoritative Guide to Advanced LTE Testing* for a detailed discussion of mobile network test configurations and KPIs. Many of the procedures and strategies outlined in that document apply equally well to virtualized networks.

4.4 Scope of Validation

A basic premise here is that NFV can largely be considered an implementation of the same functionality using new technology. As we've said, the resulting system should at the very least deliver equivalent functionality and performance as legacy systems that use dedicated hardware. Knowing the "baseline" performance and functionality of the legacy network and devices is a critical first step.

As such, the first and most important testing to conduct is essentially a regression test to validate that existing features and performance can still be guaranteed with the newly virtualized implementation of the network.

However, many of the new technologies that enable NFV and its various benefits have been proven effective for certain application areas (such as cloud storage and computing), but are not yet proven in real-time delivery networks such as the EPC and IMS core. As such, the new technologies used in virtualization should be validated specifically, in addition to the more traditional aspects listed above.

These new technologies, along with new concerns introduced by virtualization, include:

- ▶ *Dynamic scaling of capacity and performance*: The elastic properties with respect to performance are some of the most valuable benefits of NFV.
- ▶ *Orchestration and management of VNFs*: While management systems are not new, functionality is impacted by the fact that functions are virtualized.
- ▶ *Fail-over and resiliency:* The ability to spawn new VNFs as a result of detecting a fault in an existing one, as well as the ability to fall back from virtualized to legacy functions.
- ▶ *Legacy interworking:* VNFs working in conjunction with legacy elements, especially as it impacts existing functions that have to work with both virtualized and non-virtualized functions (for example, the RAN).
- ▶ *Portability:* Being able to move VNFs across multiple hypervisors, hardware hosts, etc.
- ▶ *Flexible service chaining:* The ability to define different network forwarding graphs on a service basis.

4.4.1 Managing the Migration

With NFV, there's no hard cut-over as with other infrastructure migrations. At each phase, operators must begin with baselining current functionality, then plan, test, and secure new implementations. Upon migration, performance must be validated again, then monitored on an ongoing basis to maintain quality.

"Lab-to-live" testing should encompass:

- ▶ **Planning:** Evaluating new devices and architectural models on performance criteria vital to customer satisfaction (speed, latency, reliability, overall application quality).

- ▶ **Design:** Replicating the complexities and variables of virtualized wireless networks.

- ▶ **Quality:** Measuring/guaranteeing subscriber QoE for critical applications in the face of interference, mobility, outages, heavy traffic loads, and other variables. For equipment manufacturers, this would also include validating competitive claims.

- ▶ **Security:** Replicating security attacks.

- ▶ **Deployment:** Replicating field issues.

- ▶ **Visibility:** Maintaining QoE as architectures evolve and services scale.

Regression testing is needed to ensure that existing functions are not broken during the introduction of new variables (features, hardware changes, hypervisor changes, etc.). This is a continuous process that must extend beyond migration and start-up for the lifetime of the network.

In the following chapters, we'll take a more detailed look at evaluating components of the virtualized infrastructure, then at test cases for real-world carrier network deployment scenarios.

Evaluating the NFV Infrastructure

Chapter 5: Evaluating the NFV Infrastructure

In moving forward with NFV, the virtualization infrastructure needs to be selected carefully. Many considerations come into play and each choice can significantly impact the overall strategy.

Common infrastructure elements will not only determine the performance and features of the NFV system as a whole, but potentially cause bottlenecks. For each element, specific aspects of performance must be taken into account:

5.1 Hardware

Server features and performance characteristics will vary from vendor to vendor. The obvious parameters are CPU brand and type, memory amount, etc. Additionally, support for specific software optimization APIs within the hardware can have a significant impact on performance. The performance level of NICs can make or break the entire system as well.

5.1.1 Driver-level Bottlenecks

At the server level, routine aspects such as CPU and memory read/writes can cause underlying issues. And in most cases, more than one type of server platform from more than one vendor will be in play. Testing must be conducted to ensure consistent and predictable performance across multiple platforms as VMs are deployed and moved from one type of server to another.

With physical NICs, performance can be impacted drastically by simply not having the most recent interfaces or drivers.

Virtualization Bottlenecks

- Virtual Machine (VNF), Virtual Machine (VNF) — ④ Virtual Machine Bottleneck
- ③ Communication Bottleneck — Host vs. Guest OS
- Hypervisor / Virtual Switch — ② Virtual Switch Bottleneck
- ① Driver Level Bottleneck — Server Platform

5.2 The vSwitch

There are many options and factors to consider in selecting a vSwitch, some of which come packaged with the hypervisor, while others are standalone. vSwitches vary from hypervisor to hypervisor, with some favoring proprietary technology and others leveraging open source.

Some hypervisor vendors have also designated specific functions within this layer that others have not. For example, some vSwitches provide very basic L2 bridge functionality while others act as full-blown virtual routers.

In comparing and evaluating options for unique production environments, operators will need to weigh vSwitch performance, or throughput, against resource utilization. Testing should begin by baselining I/O performance, then progress to piling virtual functions on top.

While provisioning the vSwitch, careful attention must be given to resource allocation and the tuning of the system to accommodate the intended workload (data plane, control plane, signaling). Overprovisioning must also be avoided. The virtual equivalent of "throwing bandwidth at the problem," excessive allocation of CPU, memory, and other resources not only causes waste but can actually degrade switch performance in a virtualized environment.

The software used to implement virtual switching within the NFV infrastructure is also important because the vSwitch introduces new variables, and can become a bottleneck in the system.

5.3 The Hypervisor

Moving up the stack, operators again have multiple choices in providers with both commercial and open source options available. The commercial products may have more advanced features, while the open source alternatives have the broader support of the NFV community. And while it would be ideal to consolidate to a single type, most real deployments will typically feature more than one flavor.

Hypervisors provide the ability to strictly provision virtual resources (memory, CPU, and the like) to each VM. Most can loosely provision them, enabling the ability to oversubscribe the server hardware.

They also provide the ability to start, stop, and "snapshot" a VM, which enables backup and re-provisioning, or moving from the lab to the network.

Hypervisors have common feature-sets, providing the ability to virtualize the underlying server hardware and provision VMs, but they also have unique features and performance. In selecting a provider, it's important to look at both the overall performance of each potential hypervisor, and the requirements and impact of its unique feature set. The ability of its underlying hardware layer (L1) to communicate with upper layers should also be evaluated.

5.3.1 Hypervisor-level Bottlenecks

Hypervisors, along with the VM managers, provide the ability to move a VM from one server to another (known as re-hosting) through a VM migration. Performance during a VM migration will vary; some can provide near-hitless migration while running. This "portability" feature is a key requirement of NFV since VNFs will need to be moved if they require more resources than the current server can provide, or if the server needs to be taken out of service for maintenance. Hypervisors will also vary in terms of their integration and support of automation and orchestration software.

5.4 VM Manager

This component may or may not be part of the orchestration software. One of the main decisions to be made is whether or not to use OpenStack for the inbound API. OpenStack is becoming an industry standard, and may facilitate selection of other elements, providing an open and standard API (versus a closed system).

5.5 Management and Orchestration

Here, the profound fundamental shift from managing physical boxes to managing virtualized functionality requires vastly increased automation. Some operators are even investing in developing their own orchestration system to handle provisioning and management.

Orchestration will be responsible for VM instantiation and networking configuration for all VMs, which can become critical in supporting the new benefits promised by NFV. Also, the orchestration system will communicate with the element management system of the overall network, and therefore will most likely need to support the existing system deployed in the network.

The same considerations may come into play at this level as for the VM manager. Is it an open or closed system? Are the features supported by the system sufficient, and is it able to control many types of elements via OpenStack? Is that important to the overall strategy?

5.5.1 M&O Bottlenecks

VM managers are critical to keeping performance from becoming bogged down. Communication with the orchestration layers directly impacts the ability of the system to accommodate changes and establish underlying networking connectivity on VMs themselves.

5.6 Virtual Machines (VMs)

Above the vSwitch and hypervisor, VMs themselves can impact performance. Each requires virtualized resources—such as memory, storage, and vNICs—and each involves a certain number of I/O interfaces.

In deploying a VM, it must be verified that the host OS is compatible with the hypervisor. For each VNF, operators need to know which hypervisors the VMs have been verified on, and assess the ability of the host OS to talk to both virtual I/O and the physical layer.

5.6.1 Virtual Machine Bottlenecks

A virtual function on a VM will have its own limitations depending on resource requirements and how applications are written. To troubleshoot and ensure performance, the use of a virtualized tap is critical. Virtualized monitoring solutions such as Ixia's Phantom vTaps monitor individual vNICs and connectivity from each individual VM down to the hypervisor layer, helping to isolate and validate the performance of the ultimate virtual application.

5.7 Real-world Scenarios

Once system requirements are understood, and parameters for evaluating each component or function have been defined, test methodologies can be easily repeated using either traditional or virtual test solutions. After benchmarking and baselining small-, medium-, and large-scale configurations, operators can easily re-deploy snapshots of test scenarios allocating similar resources.

Let's take a closer look now at some specific test cases for common migration scenarios.

NFV Test Cases

Chapter 6: NFV Test Cases

In this chapter, we'll take a detailed look at several use cases for validating the virtualization of critical networks and functions. We'll consider three areas of operator networks that are prime targets for virtualization:

- ▶ The Evolved Packet Core (EPC) in mobile networks
- ▶ The IP Multimedia Subsystem (IMS)
- ▶ Broadband access networks

We'll also take a close look at how virtualized network elements work in concert with legacy physical elements. The test cases described here are performed in a lab setup designed to mirror an operator's production network. Individual VNFs can be evaluated as well as multiple VNFs working together as a system under test (SUT).

6.1 Basic Test Setup

There are two basic test architectures for testing virtualized functions:

- ▶ The first uses hardware-based test systems to emulate subscribers, clients, and devices to test one or more VNFs.
- ▶ The second uses a virtualized test system. The virtual test system acts as a VM/VNF within the same or a different server as the VNF(s) under test.

Additionally, there will be instances where both physical and virtual testers are used and both physical and virtual devices are tested together.

6.1.1 Basic Test Setup Using a Physical Tester

The diagram below shows the setup for testing virtualized networks using a traditional hardware-based solution. In this example, one or more VNFs are provisioned on the server and the network is provisioned to connect to the physical NICs of the server. This enables testing with traditional equipment, in this case an Ixia test platform.

Generic Test Setup for Testing a Virtualized Network Using a Hardware-based Test Solution

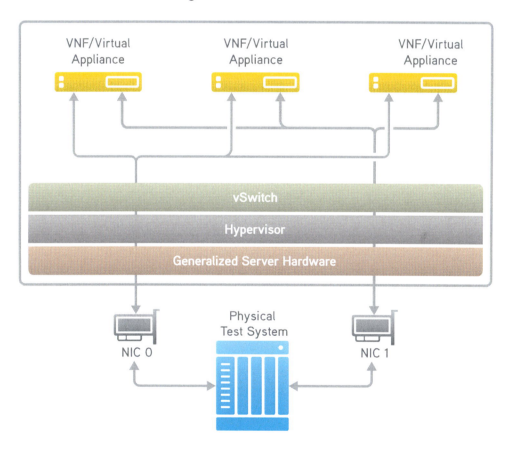

Testing of mobile VNFs mapped to physical NIC interfaces addresses many of the same aspects of performance as traditional testing including:

- Forwarding performance (loss, latency, throughput)
- Control-plane performance (calls/sec, sessions/sec, maximum concurrent calls/sessions)
- Multi-protocol/multi-dimensional testing

Unique testing and variables include:

- Performance of the vSwitch and VNF
- Determining the optimal resources (CPU/memory) allocated to the virtual appliance to meet the performance requirement
- Instantiation of a service (how fast)
- Termination of a service
- On-the-fly changes to the performance of a service (elasticity)
- Moving (VM migration) a service

6.1.2 Basic Virtualized Test Setup

The figure below shows the use of virtualized testing at a high level. Testing is inserted into the virtual environment using Ixia's IxVM virtualized test solution to provision test ports that are used to evaluate data-plane or control-plane (protocol) functionality and performance.

Generic Test Setup for Testing a Virtualized Network Using Virtualized Test Simulations

Note: Simulations may or may not reside on the same hardware platform (server)

Testing virtually by inserting test VMs into the virtualized server includes:

▶ Testing the vSwitch for performance

▶ Testing each of the virtual appliances

▶ Testing the chaining of virtual appliances

▶ Isolating and testing each function before mapping to physical interfaces

This scenario can be used to recreate various environments quickly:

▶ Virtualization technologies by their nature allow for snapshots

▶ Development, QA, and production environments can be quickly recreated through snapshots

▶ The creation and management of complete mobile networks can be automated

▶ NFV configurations can be validated without "truck rolls"

Testing can also be conducted with a combination of physical and virtual test ports. Let's take a look at how these basic methodologies and setups can be evolved to test realistic deployment scenarios.

6.2 Validating the Virtual EPC

The testing architecture shown below allows for the validation of end-to-end EPC functionality. While the architecture shown uses virtualized test components, the decision to do so can be made on a case-by-case basis as discussed above.

In validating the virtual EPC, the regression testing described earlier can be performed with high-load tests run to validate:

- ▶ *User-plane performance:* The packet-forwarding capabilities of individual VNFs.

- ▶ *Control-plane performance:* The signaling capacity of the network elements. Throughout the network, the control plane can easily become a bottleneck.

- ▶ *QoS and service validation:* This is critical. Ultimately, subscribers notice the quality of the services, and their main point of reference will be the performance of the legacy system. If degradation is noticed, it will be difficult to overcome from a satisfaction standpoint.

- ▶ *Policy and charging functions*: Billing errors represent another high-profile source of subscriber dissatisfaction that must be avoided to prevent churn.

Testing the Virtual EPC

```
┌─────────────────────────────────────┐  ┌──────────────────────┐
│  SUT                                │  │  Virtual Tester      │
│  vEPC                               │  │   ┌──────────────┐   │
│   ┌──────┐          ┌──────┐        │  │   │   eNodeB     │   │
│   │ vHSS │          │ vPCRF│        │  │   └──────────────┘   │
│   └──┬───┘          └──┬───┘        │  │   ┌──────────────┐   │
│   ┌──┴───┐  ┌──────┐ ┌─┴────┐       │  │   │   Internet   │   │
│   │ vMME │──│ vSGW │─│ vPGW │       │  │   └──────────────┘   │
│   └──────┘  └──────┘ └──────┘       │  │ Emulated Subscribers │
│                                     │  │    and Applications  │
└─────────────────────────────────────┘  └──────────────────────┘
              ↕                                      ↕
┌────────────────────────────────────────────────────────────────┐
│                           vSwitch                              │
├────────────────────────────────────────────────────────────────┤
│                          Hypervisor                            │
├────────────────────────────────────────────────────────────────┤
│                  Generalized Server Hardware                   │
└────────────────────────────────────────────────────────────────┘
```

In addition, tests can be devised to validate NVF-specific technologies and capabilities. These include:

▶ ***Elasticity of capacity and performance***. By varying the load amounts produced by the test system (sudden surges in subscribers, user plane demand, etc.), the orchestration system will be triggered to add or decrease performance and capacity dynamically.

Obviously, this should be transparent to users and the services delivered. Spikes in demand for capacity should be met immediately, without causing any noticeable performance degradation or introducing latency.

▶ ***Resiliency***. By either generating or simulating faults in the SUT, the ability of the system to diagnose and resolve those faults should be verifiable. In such scenarios, sessions that are active within the affected NFVs should be resumed as quickly as possible, and new requests serviced quasi-immediately. The amount of failed sessions or interrupted services should be fully minimized.

- **Portability.** The same suite of tests run against VNFs using a set of hypervisors and physical hardware can be run against other hypervisors and hardware. This will validate the portability of the VNFs, and also verify performance of virtualized functions against various hardware platforms.

- **Multi-tenancy.** Different functions can be instantiated on hardware platforms shared with other VNFs. For example, one hardware platform could potentially run both a PGW and a policy and charging rules function (PCRF). The ability of the NFVs to perform predictably in this case, as well as the ability to perform transparently, is verified (compared to each having a separate hardware platform).

Also, see Ixia's *The Authoritative Guide to LTE Testing* for a detailed discussion of mobile network test configurations and KPIs. Many of the procedures and strategies discussed in that document apply equally well to virtualized networks.

6.3 Testing a Network with a Combined Legacy and Virtualized EPC

This use case is important because of the simple reality that the entire network will generally not be virtualized in one shot. There are two basic cases where this can have an impact:

- When elements of the virtualized EPC will need to interact with elements of the legacy EPC

- When outside elements will need to interact with both the vEPC and the legacy EPC (prominent examples include the RAN, the billing system, the IMS, roaming partner systems)

The figure below shows the basic test setup, using physical test appliances. With the test equipment simulating both the RAN and the IMS, as well as possibly a foreign EPC, the following test cases should be attempted:

▶ Handovers between eNodeBs that are attached to the vEPC and the legacy EPC

▶ Inter-system handovers between the local system and the foreign EPC

▶ Simulation of a fault on a VNF, causing the session(s) to be transferred to the legacy EPC as a fail-over mechanism

Combined Legacy and Virtualized EPC Testing

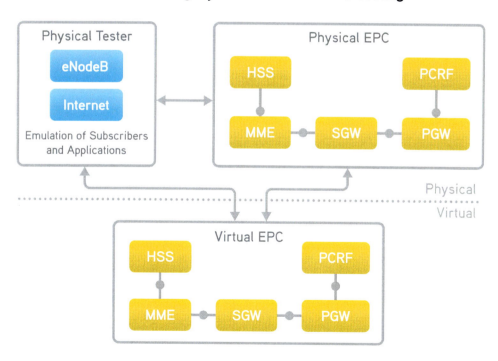

6.4 Testing Virtualized Elements Within the EPC

As we've said, virtualization of the EPC will often occur incrementally with control plane elements such as Diameter servers, the HSS, and PCRF likely to be among the first migrated. The test setup shown below can be used to validate the operation of the vHSS (virtual Home Subscriber Server) and vPCRF, which will be interacting with legacy EPC elements such as the MME and PGW.

Two different architectures are shown:

vHSS and vPRCF Tested with a Core EPC Simulation Using Hardware-based Test Equipment

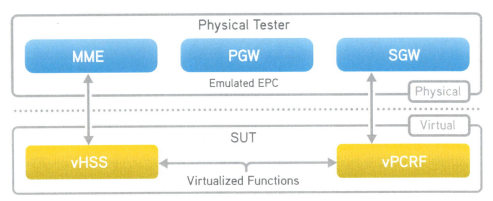

vHSS and vPCRF Tested Using Virtual Simulations

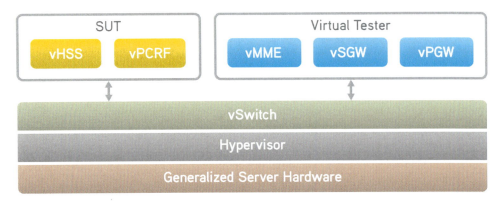

Both test architectures aid in:

- Understanding performance (transactions/second) and scale (number of subscribers, number of open transactions for the PCRF) of the devices under test (DUTs)
- Identifying bottlenecks
- Obtaining accurate dimensioning information for use in engineering the network

6.5 Testing IMS Virtualization

This use case focuses on validating the virtualization of IMS and the associated applications, such as VoLTE. Like the EPC, IMS will not be virtualized in a vacuum: the existing, legacy IMS and its services will continue to exist for a long time to come, so interaction between the two must be maintained.

Emphasis is placed on this aspect in the test architecture illustrated below which includes:

- Multiple virtualized IMS networks and multiple application servers (ASs) are placed under test. Typically, these will be in different cloud environments, because IMS is likely to be deployed in different geographic locations, creating a highly distributed configuration.
- A "real" IMS, representing the legacy IMS that exists today, along with its services.
- A simulation providing UE traffic for all applications, like VoLTE, file and video share, and also new web-based applications such as WebRTC. A hardware-based simulation is shown to illustrate the distributed nature of the UEs, which, in a real environment, would actually be connections to PGWs spread across multiple locations.

Combined Legacy and Virtualized IMS Testing

The suite of tests to be run in this case should encompass the ability to exercise and measure the following:

- **VoLTE QoS**: MOS (mean opinion score) measurements to ensure voice quality is sufficient
- **Latency**: Call setup times, call connect times, application response times
- **Capacity:** The maximum amount of calls sustained with acceptable QoS

Testing should also include quickly ramping up the number of simulated UEs, placing stress on the system in a short amount of time to verify the dynamic capacity-increase functionality of the vIMS. Operators must verify that capacity is downsized after the number of UEs is reduced.

Demand on the applications hosted in the legacy network should then be increased to verify capacity expansion into the vIMS.

6.6 Testing vBRAS Functionality

In the following example, a physical test system is connected to a virtualized Broadband Remote Access Server or vBRAS, the device under test. The tester emulates broadband access subscriber traffic (e.g. homes with DSL connections) on one side and internet services on the other. Important KPIs include subscriber capacity and subscriber setup rates (PPP session rates).

Testing Virtualized Broadband Remote Access Server (BRAS)

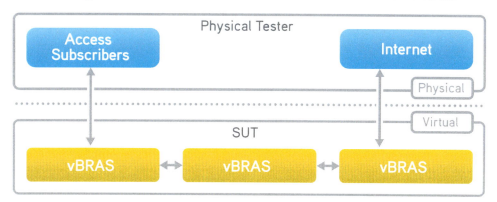

In this example, which is based on testing done by Ixia in conjunction with 6Wind, several VMs were provisioned as vBRASs to achieve the desired scale. This illustrates a major benefit of virtualization in that, if the performance limit of a device is reached, additional VMs can be deployed and the workload divided to achieve the desired scale and elasticity.

6.7 Beyond Migration

Virtualizing network functionality is just the beginning. Ensuring quality—and that the benefits of NFV and SDN continue to be realized—requires ongoing vigilance.

So let's take a look at the latest best practices in virtual visibility.

7
Maintaining Visibility

Chapter 7: Maintaining Visibility— Monitoring, Access, and Control in a Virtualized Environment

Both in the preparations leading up to deployments and on deployed networks, it is essential to maintain visibility into virtualized environments as scope and complexity increase. Ongoing network monitoring and visibility are also critical to security; specifically, maintaining the effectiveness of tools such as intrusion detection/prevention systems (IDS/IPS), data leak protection systems (DLPs), and firewalls.

New and evolving visibility challenges include:

- ▶ ***Pinpointing performance issues.*** With so much occurring within a single server, it becomes increasingly difficult to detect whether problems stem from issues at the hypervisor level, with the virtual switch, or from specific VMs. The inability to source issues to specific networks or devices can easily result in finger-pointing and wasted time.

    The challenge can be complicated by hybrid administration in which different teams or professionals have access and visibility into different elements. Greater collaboration is needed that can best be enabled through greater visibility.

- ▶ ***Blind spots.*** Gaps in visibility typically occur because traditional security and performance monitoring tools can't see above the vSwitch, where the east-west traffic flows. Inbound-outbound traffic is visible, but only represents a fraction of the overall visibility needed.

A loss of visibility into virtualized functions and traffic flow between guest VMs can occur on shared virtualized hosts, creating attractive "hangouts" for malicious intruders. Performance issues may lurk here undetected as well. Increased visibility is needed in order to detect and protect what otherwise could not be seen.

▶ ***The momentum of virtualization.*** As we've seen, the benefits of virtualization increase in proportion to how much and how quickly the effort proceeds. As virtualized environments expand and mature, it is vital that administrators have constant access to reliable information on how performance data is being used. Virtual machine "sprawl" may arise as VM creation and cloning expands rapidly, making it harder to keep track, and also to keep policies current.

With server sprawl, it is much easier to test and deploy QA and development in staging environments. As these are not productions systems, they don't have up-to-date security policies in place, and thus present another gap in system management.

▶ ***Maintaining full visibility into inter-VM –or "east-west"— traffic within servers***. Unseen inter-VM traffic on a shared server constitutes a dangerous blind spot. In a traditional environment, traffic is visible on the wire connected to the monitoring tools of choice. Inter-VM traffic, however, is managed by the hypervisor's virtual switch without traversing the physical wire that is visible to monitoring tools.

- **SPAN and tap port limitations.** Running in promiscuous mode is the virtualization equivalent of physical network SPAN ports. Operating in this mode degrades performance and opens up new potential security breaches. There's also no way to filter specific traffic—a major issue in multi-tenancy environments.

- **Maintaining compliance.** As with physical networks, access audit trails are necessary in the virtualized environment to document compliance with industry regulations, and avoid potential fines and bad publicity.

- **Load balancing.** Under virtualization, balancing loads efficiently and separating network traffic becomes more challenging. VMs move between physical servers, increasing the possibility of untrusted VMs communicating with sensitive VMs on the same virtual switch.

- **Complexity.** Multiple tools, probes, interfaces, processes, functions, and servers are involved in virtualization. Monitoring tools tasked with filtering data at rates for which they were not designed can quickly become overburdened. Network speeds are obviously dynamic (1/10/40/100Gbps), and without proper filtering of traffic of interest, oversubscription and dropped packets can occur.

As these and other factors keep actionable information from reaching the right monitoring tools, migration schedules may suffer along with performance.

7.1 Virtual Visibility: Evolving Goals and Best Practices

To cope with the new and evolving challenges introduced by NFV, visibility infrastructures must increasingly:

- Operate without negatively impacting the performance of the virtual environment

- Enable regulatory enforcement across the converged physical and virtualized infrastructures

- Integrate smoothly with virtualization technologies without requiring architectural changes or adding a large footprint

- Support the elasticity of the infrastructure and "follow" machines as they are migrated or vMotioned for optimized performance

At the most fundamental level, the goal is to achieve visibility without interference—a means of exporting traffic of interest from VMs to the appropriate monitoring tools. Several options have been proposed—adding inspection VMs, installing clients on VMs to capture and direct traffic, and others—with each introducing its own trade-offs, such as sacrificing VM or application-level vCPU, vRAM, and/or storage.

As the complexities of virtualization unfold, neither traditional taps, nor any other conventional solution appears to be able to capture all the traffic that flows between VMs. Thus, a new virtualized tap is emerging to bridge the gap.

7.2 Phantom Taps and the New Virtual Visibility Framework

Ixia has defined a Virtual Visibility Framework that eliminates blind spots, speeds application delivery, and promotes effective troubleshooting for security, application performance, and SLA fulfillment. Inter-VM traffic monitoring works to eliminate blind spots, restoring visibility lost in virtualized server infrastructures.

Virtual Visibility Architecture

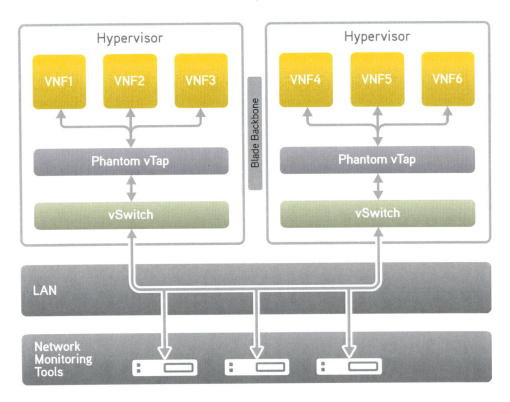

A key part of Ixia's comprehensive visibility architecture, the Phantom™ Virtualization Tap (vTap) captures the "east-west" data passing between VMs, and sends specific traffic of interest to physical or virtual monitoring tools. A software-based solution, the Phantom vTap deploys a module that resides in the hypervisor kernel, passively monitoring all inter-VM traffic. It captures only traffic of interest, without affecting production traffic.

Supporting best-of-breed hypervisors, Phantom vTaps eliminate blind spots and improve performance by working to get packets out of the virtual environment in real time with the least amount of impact on the vSwitch. The vTap mirrors packets as they pass between guest VMs with a minimum of overhead on the hypervisor level and no significant processing performed (and in turn no performance impact) on the virtual machine itself.

vTaps represent a major step towards mainstreaming virtualization, enabling rigorous network management visibility and control over sprawling virtual server infrastructures. Used with a Network Packet Broker (NPB) switch, Phantom vTaps provide the functionality achieved by traditional hardware taps and port-mirroring technologies.

Though direct connection with an NPB, packets can be sent to any existing security- or performance-monitoring tool. Smart filters can then be applied to packet streams so that only data of interest is sent to downstream management systems.

8

Ixia NFV Solutions

Chapter 8: Ixia NFV Solutions

Ixia's virtualization test and visibility solutions encompass both traditional physical test platforms and cutting-edge software-based virtualized solutions. Overall, the solution is strategically architected to cover all the bases as initiatives progress from the lab to live deployments, and to introduce procedural efficiencies (like NFV and SDN themselves).

Ixia virtualization testing assesses application and infrastructure performance, as well as visibility and optimization throughout the transition to the cloud. Advantages of this life-cycle approach include:

- Unified test applications across both physical and virtual test and visibility platforms
- Software-based traffic generation
- A virtual solution requiring no proprietary hardware
- Consistent, easy-to-use applications

8.1 Comprehensive "Real" and Virtual Test Capabilities

As we've seen, carriers have two powerful options for testing the performance of virtualized network functions:

- Using traditional hardware-based systems and mapping the service to the physical server
- Testing virtually by inserting the testing into the virtual platform

Ixia delivers both, with each approach enabling measurement of the same critical aspects of performance: forwarding rates, protocol performance, scalability, etc.

Traditional and virtual test solutions support end-to-end virtual validation of network, data center, and application performance. Product offerings include traditional hardware-based IxNetwork L2-3 and IxLoad L4-7 test solutions, along with IxVM, which provides virtual, software-based versions of the two solutions.

8.1.1 IxNetwork

Service providers worldwide rely on IxNetwork to test routers, switches, and other L 2-3 devices. Testing features high-load traffic-generation at rates up 100GE, and supports a wide variety of protocols including: IPv4/v6 routing, bridging, broadband, multicast, MPLS, Carrier Ethernet, and SDN.

8.1.2 IxLoad

Ixia's IxLoad L4-7 test solution emulates and validates the delivery of voice, video and other application traffic, as well as malicious traffic generated during security attacks. Delivering multiplay service emulation in a single application, IxLoad provides ultra-high performance and realism, including flexible subscriber-modeling in evolving service provider networks.

8.1.3 IxVM

The virtual version of IxNetwork, called IxNetwork/VM, verifies protocol functionality and validates SDN deployments. Similarly, IxLoad/VM measures application performance in virtualized network environments by providing stateful load testing of VM-based services and I/O performance testing.

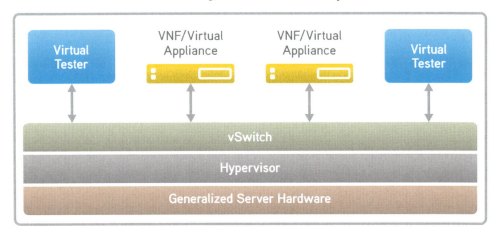

Generic Test Setup for Testing a Virtualized Network Using Virtualized Test Systems

8.1.4 Phantom Virtual Taps (vTaps)

As described earlier, Ixia's Phantom vTap provides a software-based solution that supports all leading hypervisors (VMware vSphere, Microsoft Hyper-V, Citrix XenServer) to deliver 100% visibility into virtual traffic. Featuring integration at the hypervisor kernel level, Phantom vTaps offer out-of-band monitoring that does not affect production traffic, or require any services or agents to be installed on VMs or at the application layer. The Phantom vTap is vSwitch-agnostic and can mirror all traffic within the virtual switch, apply smart filtering, and send only traffic of interest to specific monitoring tools.

8.2 The Most Trusted Names in Networking Trust Ixia

Service providers, NEMs, enterprises, and chip fabricators worldwide rely on Ixia test solutions and assessment services to validate the performance of devices, networks, services, and applications. Ixia testing serves to speed time-to-market and validate the performance of next-generation offerings, as well as the ultimate quality of the end-user experience.

We're working with carriers on front lines to:

▶ Improve network performance, resilience, security, and visibility

▶ Speed delivery of new services like VoLTE and rich-media services

▶ Implement and benefit from paradigm shifts such as HetNets, SDN, and NFV

Ixia offers the industry's only end-to-end, "lab-to-live" solution for ensuring the successful migration of critical networking functions to software- and cloud-based infrastructures. With SDN and NFV, everything old becomes new again, and we'll help ensure that it all converges seamlessly—before, during, and after virtualization.

Acronyms and Terms

AS:	Application Server
BRAS:	Broadband Remote Access Server
BSS:	Business Support System
CDN:	Content Distribution Network
COTS:	Commercial-off-the-Shelf
CSCF:	Call Session Control Function
DNS:	Domain Name System
DPI:	Deep Packet Inspection
DRA:	Diameter Routing Agent
DUT:	Device Under Test
eNodeB:	Evolved Node B
EPC:	Evolved Packet Core
ETSI:	European Telecommunications Standards Institute
HSS:	Home Subscriber Server
IaaS:	Infrastructure as a Service
IMS:	IP Multimedia System
ISG:	Industry Specification Group
IT:	Information Technology
KPI:	Key Performance Indicator
LTE:	Long-term Evolution
M&O:	Management and Orchestration
MME:	Mobility Management Entity
MOS:	Mean Opinion Score
NF:	Network Function

NFV:	Network Functions Virtualization
NFVI:	Network Functions Virtualization Infrastructure
NIC:	Network Interface Card
OCS:	Online Charging System
OFCS:	Offline Charging System
OSS:	Operations Support System
PaaS:	Platform as a Service
PCRF:	Policy and Charging Rules Function
PDN:	Packet Data Network
PoC:	Proof of Concept
PSTN:	Public Switched Telephone Network
QoE:	Quality of Experience
QoS:	Quality of Service
RAN:	Radio Access Network
SaaS:	Software as a Service
SDN:	Software Defined Network
SDN:	Serving Data Network
SLA:	Service Level Agreement
S/P-GW:	Serving and Packet Data Networks Gateway
SUT:	System Under Test
TCO:	Total Cost of Ownership
VM:	Virtual Machine
VNF:	Virtual Network Function
VoLTE:	Voice Over LTE
x-CSCF:	Call Session Control Function